Title page, opposite

MALACHITE

Description: The consistent pale green streak and the encrusting, botryoidal or stalactitic habit of this mineral reveals that it is a copper one. It belongs to the monoclinic crystal system, but is rarely found as individual crystals. Malachite is usually bright green in colour but can vary in shade. Crystals are usually transparent or translucent. A perfect cleavage is rarely seen and the ore usually exhibits an uneven fracture. It is relatively hard and can, like calcite or fluorite, be scratched with a copper coin. Malachite is a copper carbonate and is considerably lighter than the lead carbonate cerussite. Both, however, because of their carbonate content, dissolve and effervesce in hydrochloric acid. In the case of cerussite, the acid must be first warmed. Malachite is a secondary copper mineral which is mostly found in the oxidized or weathered zone of copper-rich ore bodies. It is often associated with other copper minerals, such as chalcopyrite and bornite. It is found worldwide.

The Concise Illustrated Book of

Rocks and Minerals

Richard Moody

GALLERY BOOKS
An imprint of W. H. Smith Publishers Inc.
112 Madison Avenue
New York, New York 10016

First published in the United States
of America by
GALLERY BOOKS
An imprint of W.H. Smith Publishers Inc.
112 Madison Avenue
New York, New York 10016

ISBN 0-8317-1680-0

Printed in Portugal

All photographs supplied by David Bayliss, RIDA Photo Library, with the exception of page 24 left, supplied by Ardea Photographics, London
All artworks by Bill Stone

Crystals of Fluorite

CONTENTS

INTRODUCTION

Minerals are the building blocks of all rocks. They are naturally-occurring solid substances with a definite chemical composition and a usually well-defined crystal structure. They are also characterized by specific physical properties. Minerals composed of just one naturally occurring element are termed native elements. A diamond is composed solely of carbon, albeit in its most durable and often most glamorous form. By comparison, the majority of minerals are composed of two or more elements and are termed compounds.

The physical properties of a mineral refer essentially to its shape, colour and hardness. Most minerals grow as crystals from molten liquids drawn from below the earth's crust. Each crystal is outwardly a reflection of the way in which the atoms that make up the mineral are arranged. The ultimate shape of a crystal and the perfection of its form depend on the environment in which it grows. Time and the freedom to grow within the surrounding medium allow a crystal to develop a regular shape, defined by smooth flat faces. Perfect crystals are quite rare. The most regular finds are of crystal aggregates, smaller and less perfect than the prized museum samples we admire so much.

Mineralogists and students of crystal form recognize seven main crystal systems. Each system is characterized by a given symmetry in which the regular geometric arrangement of the smooth flat crystal faces is repeated around the crystal. The variation in the number, size and shape of the crystal faces gives each crystal a characteristic 'habit'. Threadlike crystals have a fibrous habit, needlelike crystals have an acicular habit, whereas flat, blocky crystals are termed tabular. Mineral colour is extremely variable, ranging from colourless in some forms of quartz to brilliant red in rubies. Hardness also varies. Talc, for example, is easily scratched by, and thus is softer than a finger nail, whereas diamond is harder than the strongest steel.

Minerals and aggregates of minerals may be found in various geological and geographical settings. The obvious places to look are quarries, along rocky coastlines or in mountain ranges. Areas of volcanic activity, ancient or modern, are often most rewarding, as are regions where crystalline rocks have punched into the overlying layers of the earth's crust and are now exposed on the surface. Fractures and faults formed during earth movements provide pathways for mineral-rich liquids. As these liquids cool, crystals form at different temperatures and the fractures often become lined or filled with an intricate mosaic of minerals. If the right conditions prevail, excellent crystals of various minerals may be the reward of a dedicated collector.

Some rocks may be composed of a single mineral, but the vast majority are aggregates of minerals. The composition of a rock depends on its origin and all rocks belong to one of three major groups, namely igneous, metamorphic or sedimentary rocks. Rocks that crystallize from mineral-rich liquids (magmas) that ascend through the earth's crust are termed igneous rocks. They are often crystalline or glassy in texture, with the most obvious surface expression of igneous activity found in volcanic areas such as Mount St. Helens in North America or Vesuvius in Italy.

Volcanoes may be regarded as primary sources of rock production. They are often characterized by the presence of a cone, the successive layers of which correspond to individual eruptions. The rocks of the cone are composed largely of fragments of volcanic rock that have been blown skywards from a deep fissure that opens onto the earth's surface. The eruption of molten material to form a lava flow is spectacular evidence of the energy within the earth's interior. Unfortunately, lavas, particularly the top and sides, cool rapidly and crystal growth is often microscopic. However, gas spaces or vugs often form within the lava and these may become lined with calcite, sulphur or other minerals.

Where the rising magma fails to break through the earth's crust, it cools much more slowly. This may take place in a large magma chamber or in smaller magma-filled fractures. These are but two examples of the variety of ways in which igneous rocks are emplaced. The slow cooling of the magma results in crystalline rocks in which individual crystals are visible to the naked eye.

The temperature at which the magma crystallizes may vary, the end product being characterized not only by variable crystal size but also by different minerals. Granites are essentially coarse-grained rocks composed of large crystals of quartz, feldspars and mica. Quartz is a hard, glass-like mineral composed of silica (silicon dioxide). Feldspar is slightly softer and is white or pink. Micas are platy minerals and muscovite, the common mica in granites, is colourless to light coloured. Quartz, feldspar and micas, along with olivines, pyroxenes and amphiboles, all belong to the group of minerals called silicates, which is the largest of several mineral groups recognized by geologists. To classify as a silicate, a mineral must posses the elements silicon and oxygen; other elements can be present, for example, iron, magnesium and potassium.

The first igneous rocks were formed thousands of millions of years ago. As time passed, these rocks became exposed to rain, wind and ice, which breaks the rocks into fragments and transports these away from the parent rock. The accumulation of these fragments or grains results in the build up of layers of sands and gravels. These layers, when hard, are called sedimentary rocks; they are the product of erosion and formed on land or in the sea. Other sedimentary rocks are the result of the growth of reefs or the fragmentation of shells and skeletons of sea-dwelling creatures. They are termed limestones, the rocks being chemically composed mainly of calcium carbonate.

Heat and pressure on existing sedimentary and igneous rocks bring about changes in mineral composition and overall texture. The resultant 'changed' rock is termed metamorphic. Intense metamorphism can result in spectacular rocks with beautiful textures and exotic minerals.

If you plan to collect rocks and minerals, remember it can be hard and dangerous, as well as rewarding. You should always wear proper clothing and strong shoes or boots and follow these rules:-

1. Never take risks in quarries or along difficult shorelines.
2. Always obtain permission from the appropriate authorities.
3. Never damage or destroy a site simply to obtain a sample that you could purchase in a shop or obtain with less effort elsewhere, for example, from weathered material at the base of a rock face or outcrop.
4. Observe the country code.
5. Always wrap your specimens well.
6. Be well equipped as a collector with a steel hammer, hand lens, chisels, goggles and collection boxes.

Each rock in this volume is illustrated with a photomicrograph of a thin cross-section of the specimen, similar to that shown here.

AGGLOMERATE

Description: An agglomerate is a volcanic fragmental rock made up of different sized fragments which were thrown upward and outward from a volcanic vent during a volcanic explosion. The size and variety of the fragments may vary considerably. Usually the fragments are 20–30 mm (0.75–1.2 in) in diameter, but much larger pieces of rock may be observed. The individual fragments may be derived from several volcanic events, as the explosion may cut through successive layers of a cone. Fragments of the rocks that line the fissure through which the volcanic magma rises to the surface and surface rocks themselves may be included in the final product. In more recent areas of volcanic activity, the agglomerate may be rather loose and rubbly in character, but usually the fragments are bound together either by fine debris or by lava. Agglomerates are usually associated with the area close to the fissure through which the magma rises and with the hole or vent from which the magma emerges. Finer volcanic debris such as dust or ash may be carried great distances from the volcano. Blown into the atmosphere, the debris is transported on the prevailing winds. Large scale explosions such as those at Mount St. Helens or Krakatoa produced tremendous quantities of ash, some of which circled the globe and in the case of Krakatoa, remained airborne in the stratosphere producing spectacular sunsets and 'blue moons'.

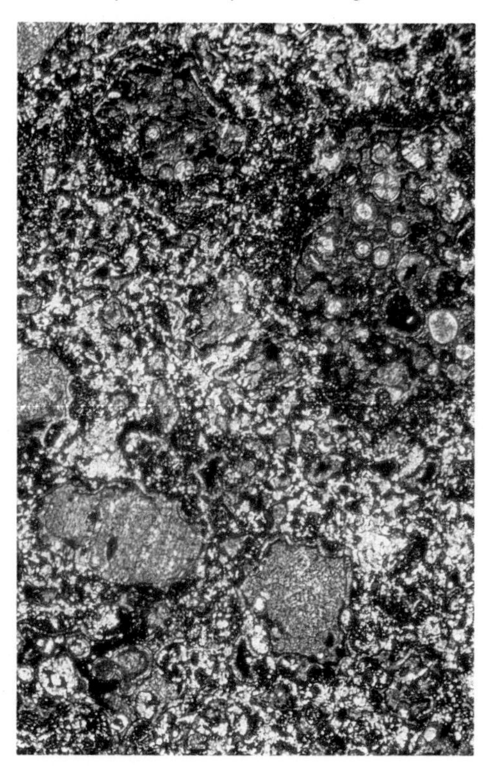

General remarks: Agglomerates, like many other volcanic rocks, are used for road building or building blocks. Some may be polished and used as facing stones.

Classification: Igneous, volcanic rock. Coarse grained product of volcanic explosion.
Locations: In areas of ancient and recent volcanic activity throughout the world.
Importance/Commercial use: Bulk resource used in construction industry.

ANDESITE

Description: These fine-grained crystalline igneous rocks are commonly found as lava flows or as bands that cut through the country rock at an angle. The correct name for such bands are dykes (or dikes in North America). Andesites are one of the most common types of volcanic igneous rock. The most important mineral in andesites is plagioclase feldspar, some of which are large enough to be seen with the naked eye. This mineral may form 70 per cent of the rock with hornblende, biotite and olivine as subsidiary minerals. Andesites are extremely varied in colour with white, grey, green, brown red, purple and black varieties recorded worldwide. Quartz forms only a minor component within these rocks; the reduced importance of this mineral indicates the rocks can be classified as intermediate rather than acidic. In hand specimen, some andesites are rather uninteresting to look at, but others have large crystals of specific minerals scattered throughout the fine-grained groundmass. Such large crystals are termed phenocrysts. Hornblende is often represented by these larger crystals. Some andesites are pocked by round holes or vesicles. These were filled by gas within the original magma. Andesites weather easily and good samples are mostly found in working quarries. When weathered, they often appear dull green due to the development of chlorite as a replacement mineral. Under the microscope, the contrast between the small crystals of the groundmass and the larger hornblendes and feldspars is easily determined.

General remarks: Andesites are quarried as a bulk resource for roads. Because they weather easily, they make poor facing or ornamental stones.

Classification: Fine-grained intermediate igneous rock of mainly volcanic origin.

Locations: In volcanic centres worldwide. Excellent outcrops occur in the Cascade Range and the Lassen Volcanic park of North America, the English Lake District and the Andes of South America.

Importance/Commercial use: Bulk resource, but second choice in areas where basalts are present.

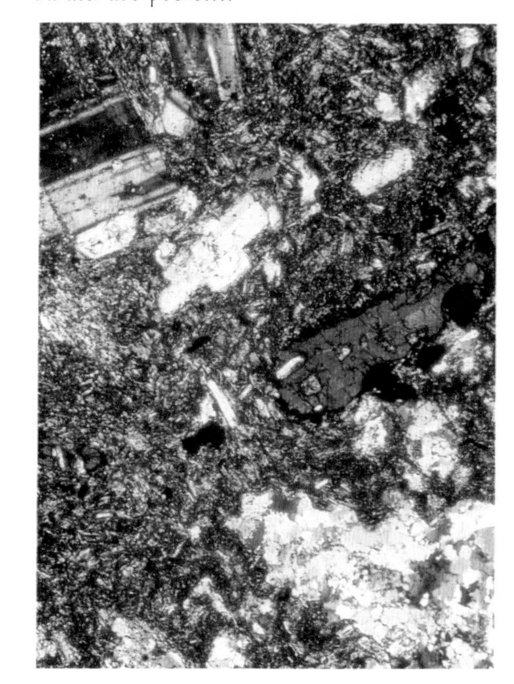

BASALT

Description: Basalts are the most common type of volcanic rock. They are fine-grained crystalline rocks extruded from volcanic vents as lava flows. Most basalts are black or grey black in colour, although some have a green or red hue. Large crystals are the exception in these rocks, but olivine, pyroxene and hornblende may sometimes be visible to the naked eye. Plagioclase feldspar in the form of anorthite and/or labradorite is the main constituent. Pyroxene and possibly olivine can also be important minerals. If labradorite is the main plagioclase feldspar, then the basalt is called a tholeiite. Basalts contain less than 52 per cent silica by volume and are termed basic igneous rocks. No quartz is present. Basaltic lavas are extremely hot (approximate 1100°C or 2000°F) and they flow rapidly outwards from the parent fissure. Basalts occur as thick flows in many areas of the world. They form plateaux lands in the Deccan region of India. Sheet flows of basalt give rise to the layered landscape of the Columbia River area of the United States. The cooling of basalt flows often gives rise to columnar jointing, spectacular examples of which occur on the Giants Causeway, Antrim, in Northern Ireland, Skye and in the French Auvergne.

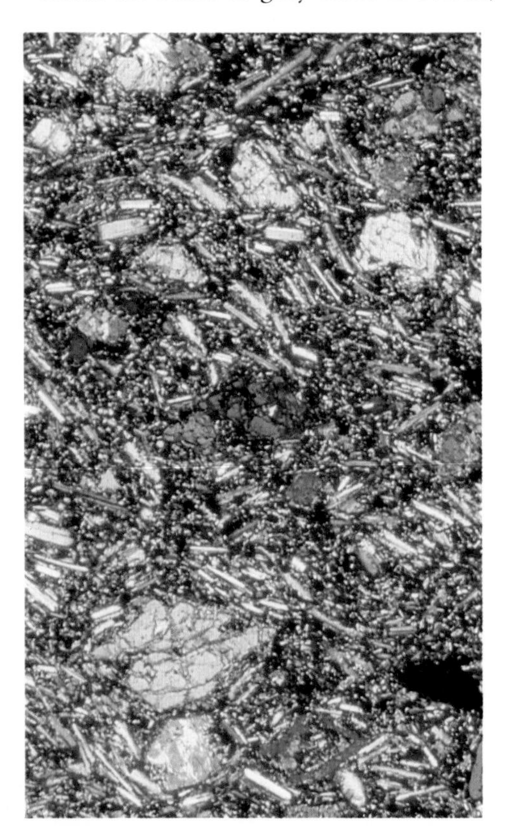

General remarks: Although many basalts are simply fine-grained igneous rocks, the occurrence of mineral-filled gas holes or vesicles can result in a rock with a beautiful overall speckled appearance.
Classification: Fine grained, basic igneous rocks of volcanic origin.

Basalt flow in Iceland.

Locations: Worldwide in regions of ancient and recent volcanic activity. Excellent outcrops in New Jersey and in south-west Scotland.
Importance/Commercial use: Important bulk resource for construction industry. Often used as facing stone and as gravestones. Used as building stone in the French Auvergne.

Description: Like agglomerates, breccias are fragmental rocks. They are however, sedimentary and are generally the product of erosion rather than volcanic activity. Breccias are composed of angular fragments which range in size from millimetres to tens of centimetres. Large fragments are set in a finer-grained matrix. The variety of the fragments depends on the place of origin, bearing in mind that the angularity of the fragments indicates they have not been carried far from their source or parent rocks. Many breccias form at the base of cliffs or steep slopes in mountainous areas. In a volcanic area, the fragments may be derived from a single rock type, whereas layered sedimentary sequences may produce a rock composed of pieces of different rocks. Transport by water would result in the corners of each fragment becoming rounded and the greater the roundness the more mature the clast, clast being the term applied to a piece of rock broken from a larger mass. The term bioclast is used to describe fragments of fossil debris. Rocks composed of large, rounded clasts set in a finer-grained matrix are called conglomerates. Another type of breccia, a fault breccia, is the product of two rock masses moving against each other. Fragments of both rock masses are broken off and fall into the plane of movement of the fault. They are then cemented together by minerals such as quartz or calcite.

General remarks: It is possible to study the processes that influence the development of breccias in most mountainous areas. Frost in a freeze-thaw regime will cause the fragmentation of the parent rock, the derived fragments tumbling downslope to form a new deposit in the form of a scree slope.

Classification: Coarse grained sedimentary rock composed of angular fragments or clasts – sediment also known as clastic rock.

Locations: Worldwide and even on the moon, where meteorite impact welds fragments into a cohesive rock.

Importance/Commercial use: Sometimes as a bulk resource for construction industry and as an ornamental stone.

CONGLOMERATE

Description: Like breccias, conglomerates are clastic rocks, the products of erosion. In contrast, the particles and clasts in a conglomerate are rounded. This is an indication of a mature sediment, one in which the original fragments have undergone attrition and have been worn down during long periods of transport. As this process continues, especially in a stream or river environment, the pebbles or boulders become more and more rounded and polished. During transport, the size of the transported particles gradually becomes smaller. It is possible to find conglomerates with clasts of one rock type, but most contain a variety of rocks and minerals picked up along the course of the river or along the length of a beach. In many localities, conglomerates will be found to rest on the eroded surface of the country rock. In this case, they are termed basal conglomerates and usually mark the advance of the sea across land, following a rise in sea level. As with breccias, the rounded pebbles and boulders of a conglomerate are often set in a finer-grained matrix. A mud matrix may indicate mass movement, either downhill during torrential rains, or the sliding of sediments over the edge of the continental margin downslope into a deeper basinal environment. Coarser sandy or gritty matrices suggest shallower water and higher energy environments, such as streams or beaches.

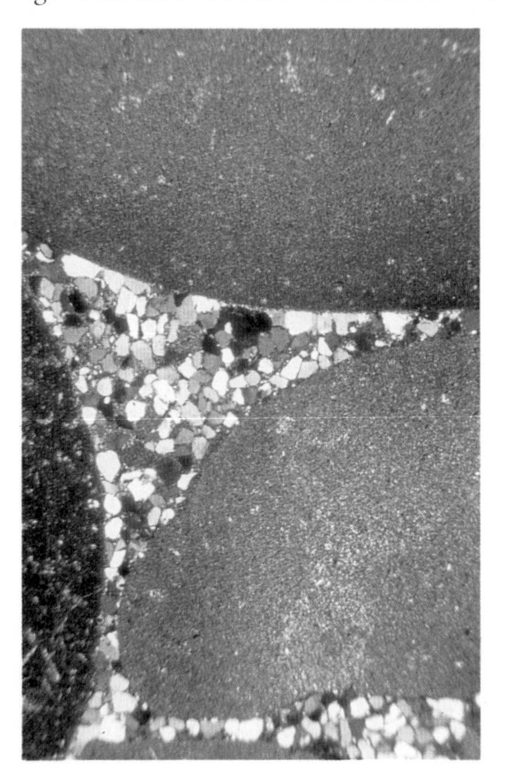

General remarks: Conglomeratic deposits can range in thickness from a few centimetres to hundreds of metres. The particles or clasts are greater than 2 mm (0.78 in) in size and may reach tens of metres in diameter.

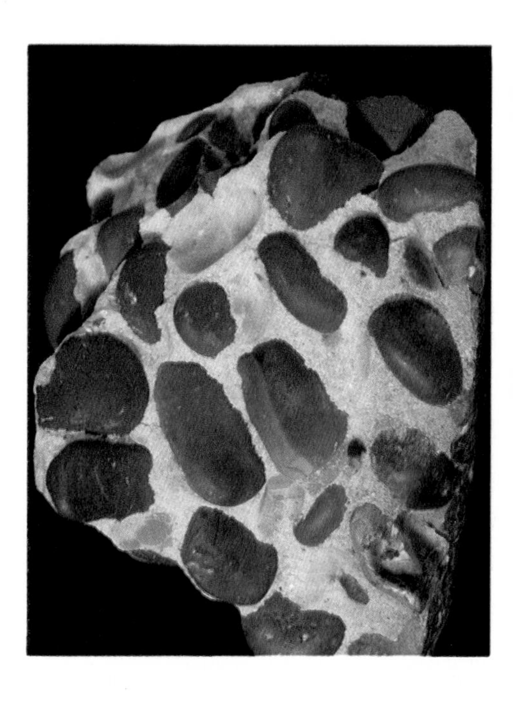

Classification: Clastic sedimentary rock. Coarse grained with clasts greater than 2 mm (0.78 in). Often termed rudaceous due to large size of grains.
Locations: Worldwide.
Importance/Commercial use: Unconsolidated materials used in concrete foundations. Important bulk resource. Ancient materials quarried or cut for facing stones.

Description A dolerite is a medium-grained igneous rock known as diabase in America; European geologists use this name to describe a metamorphosed dolerite. Most dolerites have a uniform crystalline texture. The main minerals are plagioclase feldspar, pyroxene and olivine. Silica, in the form of quartz, is absent or insignificant, and, if present, is not visible to the naked eye. The indication is that dolerites are formed of high temperature minerals. This shows that as a magma cools, different rock types are formed during different phases of cooling. It is also true that concentrations of high temperature minerals invariably result in rocks with little quartz and which are of a basic, rather than acidic, nature. Most dolerites are dark in colour, but the presence of moderately-sized plagioclase crystals may give the rock a black and white speckled appearance. Under the microscope, the most noticeable minerals are pyroxene and plagioclase. In some dolerites, the plagioclase crystals are enclosed by the larger crystals of pyroxene. This commonly seen texture is termed ophitic. Dolerites usually occur as dykes or sills. This last is a horizontal or concordant igneous intrusion, with the magma emplaced between layers of country rock. Swarms of dolerite dykes are recorded from the west coast of Scotland. The doleritic Whin Sill in the north of England is a spectacular intrusion that sometimes carries the remains of Hadrian's Wall. In the U.S.A., the Palisades Sill of New York States is doleritic, as is the Mount Wellington Sill in Tasmania.

General remarks: Outcrops of doleritic rocks are usually rounded by weathering. They are also well-jointed which increases the effectiveness of natural agents such as frost, ice and rain on the weathering of the rock.

Classification: Granular, medium-grained basic igneous rock. Low silica content.

Locations: Worldwide.

Importance/Commercial use: Bulk resource for building and construction industry.

GABBRO

Description: Coarse to medium-grained igneous rocks, gabbros are characterized by dark-coloured minerals such as augite, olivene, hornblende and biotite. Plagioclase feldspars are the most common light-coloured mineral in these basic rocks. Quartz and orthoclase feldspar may occur rarely, but as in most igneous rocks, the combination of quartz and olivine is almost unknown. As coarse-grained crystalline rocks, gabbros consist of interlocking crystals with poorly-defined shapes. The texture of a gabbroic rock is the result of the slow cooling at depth of the parent magma. Under the microscope, it is noticeable that other textures occur. In some gabbros, the larger crystals may completely enclose or surround several smaller crystals. This is termed a poikilitic texture. In so-called 'troutstone' gabbros, the plagioclase and olivine crystals are approximately the same size. Slow cooling, basic magmas often give rise to banded or layered rocks. This is the result of minerals crystallizing at different temperatures, with the denser, earlier-formed crystals sinking to a given level in the magma body. Olivine and pyroxene crystals are denser than those of plagioclase feldspar and, as a result, a colour banding is characteristic of specific gabbros. Banded gabbros occur in the Palisades Sill, New York and at St. Peter Port on Guernsey in the Channel Islands. Gabbros often outcrop as hills or mountains. They are usually associated with saucer and inverted saucer-shaped intrusions termed lopoliths and laccoliths, respectively.

General remarks: Gabbros are known throughout the world. They are associated with the mineral-rich areas of the veldt in South Africa and the Stillwater complex of Montana, U.S.A.

Classification: Coarse to medium-grained, basic crystalline igneous rock.
Locations: Worldwide with Skaergaard, East Greenland, the Lizard complex in Cornwall, England and San Marcos, California, as well-known examples.
Importance/Commercial use: Gabroic intrusions are often host to a variety of commercially valuable minerals such as chromite.

Description: In contrast to gabbroic rocks, granites are usually light-coloured. This is due to a high percentage of feldspars and to over 10 per cent of visible quartz. The feldspars include orthoclase and plagioclase which are large white or pink crystals and which characterize the coarser-grained varieties of crystalline acidic rocks. Biotite and hornblende, which are dark minerals, give many granites a speckled appearance. White, flaky muscovite is a common constituent of granitic rocks. Some granites show segregation of minerals into bands, while flow structures formed within the magma are quite commonly observed, particularly in the fine-grained acidic igneous rocks such as rhyolite. However, acidic magmas usually crystallize out over a small temperature range and granite intrusions often have a uniform texture. Large intrusions or batholiths of granite are associated with mountain-building movements. Granitic rocks are often well-jointed and weathering may result in the formation of tors. These are upstanding structures which form bare hillsides about the size of a house in moorland areas such as Dartmoor in Devon, England. The upper blocks of a tor are small and more rounded than those at the base, the topmost joints responding most to weathering. The area around each tor is often marked by scattered boulders or clitter. Granitic rocks may also occur as dykes or sills.

General remarks: Granitic rocks are among the most common igneous rocks. Their coarse-grained texture, durable nature and the growth of large, well-formed crystals makes them of interest to both amateur and professional geologists.

Classification: Coarse-grained, crystalline acidic igneous rock, with interlocking texture.

Locations: Worldwide, with notable intrusions in the Andes of South America, Devon and Cornwall, England, and in the Sierra Nevada Range, U.S.A.

Importance/Commercial use: As bulk resource, facing stone, kerb stones and as host rock for minerals such as tin and copper.

GNEISS

Description: As with igneous and sedimentary rocks, it is possible to subdivide metamorphic rocks into coarse, medium and fine-grained varieties. Gneisses are coarse-grained rocks formed by the action of heat and/or pressure on either igneous or sedimentary rocks. The processes involved bring about changes in the mineral composition of the original rock. In the case of gneiss, the changes are dramatic and are the result of high temperature and pressure regimes. Those that result from the alteration of igneous rocks (e.g. granites) are called orthogneisses, whereas paragneiss is the term used to describe such coarse-grained rocks derived from sedimentary rocks. Gneisses are foliated rocks in which the minerals occur in distinct bands or layers. In most samples, the dark bands correspond with the distribution of biotite and hornblende, the lighter layers with quartz and feldspar. Eyelike or lens-shaped clusters (augen) of quartz and feldspar are characteristic of an augen gneiss; less common minerals include kyanite, sillimanite and garnet. Gneisses are usually hard and resistant rocks. They are associated with mountain building movements and many outcrops occur high up in the Alps, Andes, Himalayas and Rocky Mountain chains. Garnet-rich gneisses are among the most attractive of all rocks. Polished gneisses are found as facing stones on many major buildings in the great capitals of the world.

General remarks: Regional pressures can result in highly-folded patterns within gneissose rocks. These attractive rocks often outcrop along rugged coastal stretches and in glaciated areas where weathering gives rise to excellent sections.

Classification: Coarse-grained, foliated, metamorphic rock.
Locations: Worldwide, commonplace in ancient continental massifs such as the Canadian Shield and Africa and in more recent mountain belts such as the Alps.
Importance/Commercial use: Used frequently as polished facing stones in the building industry. Garnetiferous gneisses yield semi-precious stones.

IRONSTONE

Description: Ironstone is a term usually applied to sedimentary rocks composed mostly of compounds of iron. Several types of ironstone are known to geologists. Sedimentary ironstones are one of the most important sources of iron. Many were formed thousands of millions of years ago during the Precambrian Era. It is thought that the oxygen level in the atmosphere was lower during Precambrian times than they are now. This allowed iron, in soluble form, to be transported and deposited in shallow sea environments. Thin layers of ironstone built up with alternations of silica to assume a banded appearance. The main iron compound in these sediments is haematite, which has a dull red colour. Banded ironstones are essentially chemical precipitates. Blackband ironstones are a mixture of organic coal-like material and iron carbonate. Another organically-formed iron ore is termed bog iron. It is usually quite pure with low sulphur and phosphorus levels. Bog iron is produced by bacterial activity in the lower levels of peat bogs. High grade iron ores are also formed in mineral veins associated with volcanic and metamorphic areas. These are usually composed of sulphides of iron such as iron pyrites. At the present day, ironstones composed of small spheres of chamosite are forming in Lake Chad, central Africa. The spheres are termed ooids.

General remarks: Ironstones are commonly found in humid tropical areas and in areas such as the Sahel in West Africa, where they form a hard crust (duricrust) over soils and rocks alike. Nodular ironstones are found in association with coal seams or black shales.

Classification: Chemical or organic sedimentary rocks.

Locations: Worldwide throughout the geological column. Banded ores found in Mesabi, Minnesota, North America.

Importance/Commercial use: Banded iron ores and, to a lesser extent, bog ores are a source of important economic minerals. Laterites are used for road building and brick making in tropical countries.

LIMESTONE

Description: Calcium carbonate is the major chemical constituent of limestones. These sediments are commonly known as carbonates and are formed either by chemical precipitation, organic activity or through the accumulation of fragments of shell material. Chalk is one of the purest forms of limestone; it is composed of the calcareous skeletons and fragments of microscopic plants and animals. Limestones can be marine or freshwater in origin. In open marine waters, beyond the continental shelf, limestones are often formed by the accumulation of myriads of skeletons of tiny planktonic organisms, which float on the surface of the oceans. At depths of 4 or 5 kilometres (2.5–3.1 miles) below the ocean surface, calcareous skeletons dissolve and only the silica-rich tests of creatures called radiolaria persist. Reefal limestones are composed of the skeletons of corals, bryozoa and coralline algae. Fragmental and detrital limestones may accumulate as a result of the erosion of reefs or of shell beds. In the warm waters of the Arabian Gulf or around the Bahamas, carbonate 'sands' composed of small (0.2 mm-1.0 mm (0.007–0.03 in)) spheres of calcium carbonate, called ooids, accumulate to form distinct shoals and banks. Ooids have an onion-like structure with successive layers of calcium carbonate formed around a nucleus. A limestone composed mostly of ooids is termed an oolite.

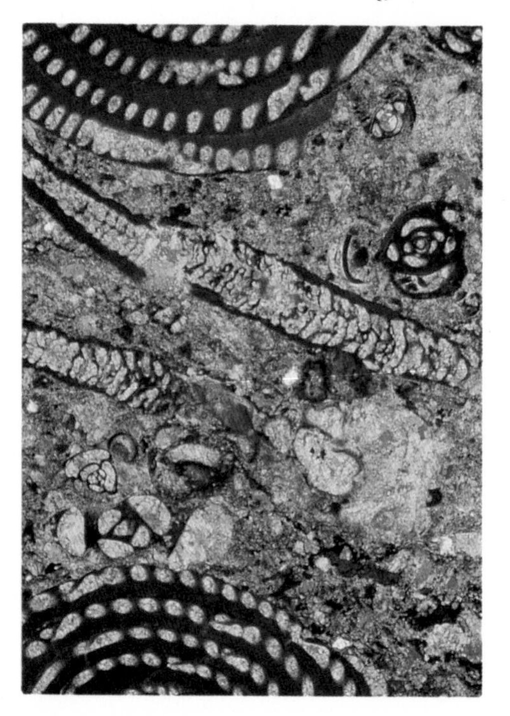

General remarks: Reefal limestones can form distinct topographic features. The Capitan reef of West Texas is a classic example of a fossil reef structure and forms the main ridge of the Guadalupe Mountains. Organic-rich limestones are main source rocks for gas and petroleum.

Lower Jurassic limestone ammonites.

Classification: Carbonate sedimentary rocks. Calcium carbonate comprises over 50 per cent of the rock by weight.
Locations: Worldwide. The Devonian, Carboniferous and Jurassic periods are noted for the deposition of reefal limestones.
Importance/Commercial use: Bulk resource for production of agricultural lime, cement for building industry, facing blocks for construction industry.

Description: A marble is a metamorphosed limestone in which the original sedimentary texture has been changed, by recrystallization, to a metamorphic texture of interlocking crystals. The calcium carbonate fossils in the limestone have become recrystallized to crystals of calcite. The main agent of change is heat or thermal metamorphism. The initial limestones are subjected to relatively low temperature changes but the degree of alteration usually obliterates the original texture of the parent rock. Marbles formed from pure limestones have an even-grained texture. Others, of which the parent rock contained impurities, may be medium to coarse-grained with a strong coloration. This is true of forsterite marble in which metamorphism has resulted in the production of a magnesium-rich variety of olivene (forsterite) and a pale grey colouration. Under the microscope, pure marbles exhibit an interlocking granular texture, mostly of calcite or dolomite crystals. The term marble is used loosely by quarrymen to describe any decorative stone that will take a high polish and is locally applied to unaltered limestones. True marbles are much prized as decorative stone and for use in sculpture. Carrara in the Liguria Province of Italy is famed for the occurrence of various marbles. The most pure were used by Michelangelo during the 16th century. During Roman times, whole columns were carved out of marble quarries for use in temples or capital buildings.

General remarks: Marble is associated with mountain building areas or areas of regional metamorphism. The rock is usually exploited as large blocks which are cut away from the quarry face by wire or steel rope. The blocks are cut and polished for use by the building industry.

Classification: Fine to coarse-grained metamorphic rocks/granular textures.

Locations: Worldwide. Best known examples outcrop in the Predazzo region of Italy, in the Three Rivers area of the Sierra Nevada and in the Adirondacks of New York State, U.S.A.

Importance/Commercial use: One of the most valued rocks known to man. Fine grained, white, even textured varieties used by sculpters throughout the ages. Important facing and decorative stone.

METEORITE

Description: As extraterrestrial bodies, it is perhaps wrong to describe meteorites as rocks. They are, however, associations of minerals and are solid in form. The majority consist of minerals such as olivene, pyroxene and feldspar or metals such as iron and nickel. Meteorites are classified on their mineral composition with the percentage of metals and other minerals being important factors in their classification. Stony meteorites are made up mainly of olivene, pyroxene and feldspare whereas a siderolite is 90 per cent iron. Glassy meteorites are called tektites. Most meteorites are small fragments of larger bodies that broke up on entering the Earth's atmosphere. Evidence suggests that in the past, huge meteorites have hit the surface of our planet. The tell-tale evidence exists in the form of impact craters. These are broad, shallow and flat-floored features. One of the largest craters is found near Winslow, Arizona, USA. It is nearly 1500 m (4,920 ft) across and 180 m (590 ft) deep. The crater was formed approximately 10,000 years ago, about the time Homo sapiens first appeared. Older craters are poorly preserved. They are obscured by weathering and the meteorites broken into millions of tiny fragments. The surrounding country rocks may be altered or metamorphosed by the extreme heat and pressure generated by meteorite impacts.

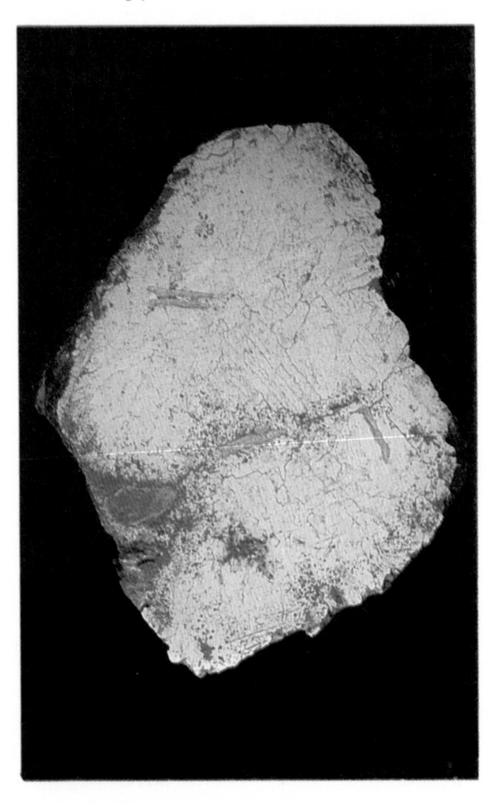

General remarks: Fortunately, meteorites of any size rarely penetrate the Earth's atmosphere. The moon is much more susceptible to meteorite impact, hence the thousands of craters that pock its surface.

Classification: Silicate or metal-rich solid of extraterrestrial origin.

Locations: Worldwide, but most commonly found in open tracts of land such as deserts.

Importance/Commercial use: Little or no commercial value except to collectors. Important evidence of extraterrestrial materials and for dating the universe.

Description: Muds are fine-grained deposits in which the grains are usually less than 0.0039 mm (0.00015 in). They may be deposited in marine or non-marine environments, but are usually associated with low energy conditions. Lakes, lagoons, estuaries and tidal flats are areas where the deposition of muds is likely. Burial and compaction causes the loss of water and results in the formation of a mudrock or mudstone. Mudrocks are also described as being 'argillaceous'. They contain ground rock materials and a variety of clay minerals. Mudstones are finer grained than siltstones in which the particle size range from 0.0039 mm (0.00015 in) to 0.06 mm (0.0023 in). Quartz and iron constitute a higher percentage of the overall composition of siltstones. Mudrocks may be fossiliferous or reworked by the activity of animals living on or in the mud before it became lithified. Body fossils and traces may be used in the recognition of the environment of deposition. In contrast to a shale, a mudstone is not a fissile rock, i.e., one which can be easily split along closely-spaced bedding planes. Mudstones tend to be medium to thickly bedded and have a subconchoidal or blocky fracture. Clays are composed flaky clay minerals, are plastic when wet and poorly bedded. Black shales deposited under anoxic (oxygen-poor) conditions may be rich in carbonaceous matter and pyrite.

General remarks: Mudrock is a term used to describe the finer-grained clastic rocks. They are argillaceous or clay-rich.

Classification: Argillaceous, clastic sediments.

Locations: Worldwide since the Precambrian.

Importance/Commercial use: Specific sediments used in chemical and pharmaceutical industries.

OBSIDIAN

Description: Also known as pitchstone, obsidian is a black shiny volcanic rock. It is essentially a volcanic glass which has no crystal structure. The term used to describe such rocks is amorphous. Glasses, such as obsidian, are formed when a granite magma cools almost instantaneously. The glassy or hyaline appearance of obsidian is emphasized by the presence of conchoidal fractures. The rock has a similar chemical composition to the acidic rock rhyolite. Some glasses exhibit small spheres or embryonic crystals which may show up as light-coloured blebs or needles within the black matrix. These structures are the result of the growth of quartz or alkali feldspar crystals. The process is termed devitrification and in time the rock may become completely crystalline. Plagioclase crystals, in radial or sheaf-like aggregates, may develop in basaltic glasses. The pale crystalline patches stand out against the dark vitreous, matrix; rocks possessing this appearance are called 'snowflake' obsidians and are attractive collectors items. Obsidian was fashioned by primitive peoples into arrowheads or into tools for cutting meat and cleaning skins. Volcanic glass is usually found associated with lava flows or in dykes. The most common association is with rhyolite, which is not surprising as obsidian is the chemical equivalent of rhyolite.

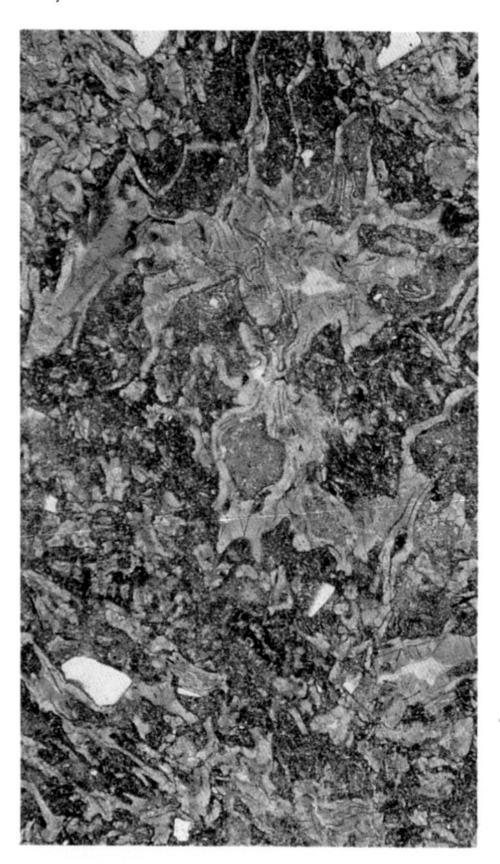

General remarks: Obsidian is an attractive collectors item and was once used as a raw material for the production of rock wool.

'Snowflake' obsidian.

Classification: Amorphous acid igneous rock.
Locations: Worldwide in volcanic provinces.
Importance/Commercial use: Of no commercial value.

Description: Phosphate is a chemical compound which is found in bone as calcium phosphate and magnesium phosphate. It was first extracted, according to scientific history, from human urine in 1669 by the German alchemist Henning Brandt. Phosphate is essential to life. In sedimentary rocks, phosphate is present in a variety of apatite called collophane. Collophane is described as cryptocrystalline as it is impossible to see individual crystals without the use of a high-powered microscope. Collophane usually occurs as ooids or pellets or as the remains of vertebrate animals such as fish. In marine deposits, the droppings of fish and other animals are major constituents of many phosphatic sediments or phosphorite. Nodules of calcium phosphate can form, by precipitation, in offshore areas where waters rich in organic material rise up from great depths in the ocean basin. This upwelling is commonplace on the western coastline of the Americas and is particularly important off the shores of Peru. The same processes undoubtedly took place 60 million years ago when possibly the richest phosphate deposits yet recorded were deposited along the coastline of what is now northwest Africa and north Africa. The phosphatic rocks of Morocco and Tunisia are rich in vertebrate remains. They also contain an abundance of animal droppings termed faecal pellets and coprolites. Phosphate sediments are usually reworked by water currents and tidal waters and a concentration of larger pellets often occurs.
General remarks: Phosphate contains a relatively high percentage of radioactive material. Petroleum geologists can recognize the presence of phosphate rock by a marked signature on their gamma logs.
Classification: Sedimentary rock of chemical and organic origin.
Locations: Worldwide, but most-noted deposits are in Morocco, Spanish Morocco and Tunisia.
Importance/Commercial use: Source of fertilizer and 'super phosphate'.

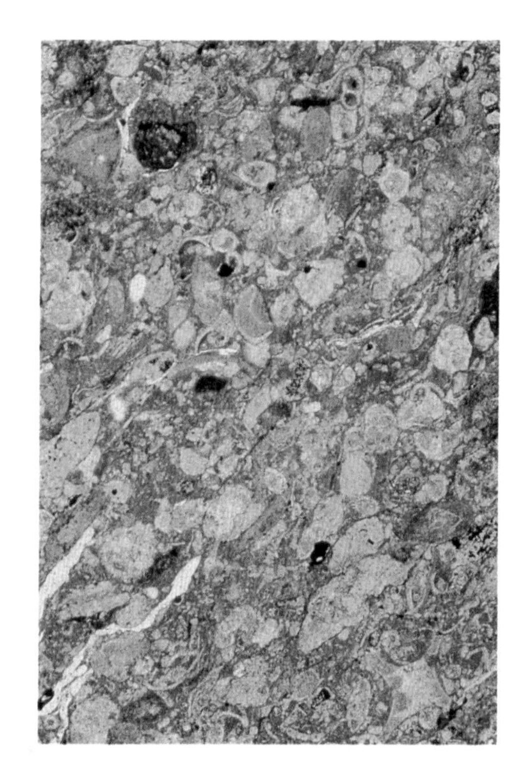

PUMICE

Description: During the initial gas-rich phases of a volcanic eruption, lava expelled from the neck of the volcano may develop a frothy glass-rich rock over its upper surface. The density of the parent material is greatly reduced by the presence of numerous holes or vesicles, which represents a solidified lava having numerous gas bubbles. Chemically, the resultant rock has the same composition as the parent lava. Pumice is an acidic rock closely related to rhyolite and obsidian. In contrast to these rocks however, it is extremely light and can float on water. Pumice is also carried skywards during explosive episodes ultimately falling back to earth. In time, layers of pumice accumulate, forming pyroclastic deposits. Thick layers of pumice are known from the volcanic provinces of northern Morocco, amongst others. Air fall deposits are often a mixture of pumice, blocks of lava, country rock, volcanic bombs and fine volcanic dust. Pumice is usually white in colour but shades of yellow, red and green are known. Crystals of quartz, sanidine (alkali feldspar) and plagioclase feldspar may occur in the fine-grained matrix. In some areas, the vesicles may be filled by the growth of late stage minerals such as zeolites. Numerous vesicles may occur in basaltic lavas; the descriptive term applied to these denser rocks, in which the vesicles are filled by minerals, is amygdaloidal. Vesicular, clinkery, basaltic material is often described under the name scoria. This material is not as light as pumice and is often dark-coloured.

General remarks: Pumice is used as an abrasive and pumice stones are sold as a cosmetic aid to remove rough skin.
Classification: Acid igneous rock.
Locations: Worlwide in most volcanic provinces.
Importance/Commercial use: As an abrasive.

Description: The term pyroclastic describes materials produced during a volcanic explosion. In recent times, the destruction of Mount St. Helens (USA) was spectacular evidence of the great volumes of material produced during an eruption. In contrast to the liquid nature of lava flows, pyroclastic material falls as dust and volcanic bombs. Such debris is usually associated with cone-shaped volcanoes. These are steep-sided and are the product of the violent emission of viscous, acidic magmas. The fragmental material may be solid or 'wet' at the time of the eruption. Fine ash is thrown up into the atmosphere, whereas the heavier fragments fall back to earth. The general term for solid material thrown out of the volcanic vent is ejectamenta. Geologists classify airfall products on grain size. Pyroclastic rocks can range from a fine-grained dust, in which the particles are less than 0.06 mm (0.002 in) in diameter, to coarse grained fragmental rocks in which individual clasts exceed 20 mm (0.78 in). Volcanic ash contains particles between 0.06 mm (0.002 in) and 4 mm (0.15 in) in size. A consolidated ash is termed a tuff. Pyroclastic rocks may contain shards of volcanic glass and may even be fossiliferous, if the pyroclastic material settled out to the sea, (e.g., Vesuvius is only a relatively short distance from the sea). The fossils are the remains of animals, and plants trapped down below the dust and debris. The mummified remains of the citizens of Pompeii, Italy, are an example of this process of preservation.

General remarks: Pyroclastic rocks are the main constituents of volcanic cones. When deposited in water, the volcanic debris will behave like sand or clay particles and exhibit depositional characteristics typical of lake, river or marine environments.

Classification: Fragmental volcanic rock.
Locations: Worldwide. Throughout the geological column. Useful markers for correlating rock sequences.
Importance/Commercial use: Bulk resource, building stone.

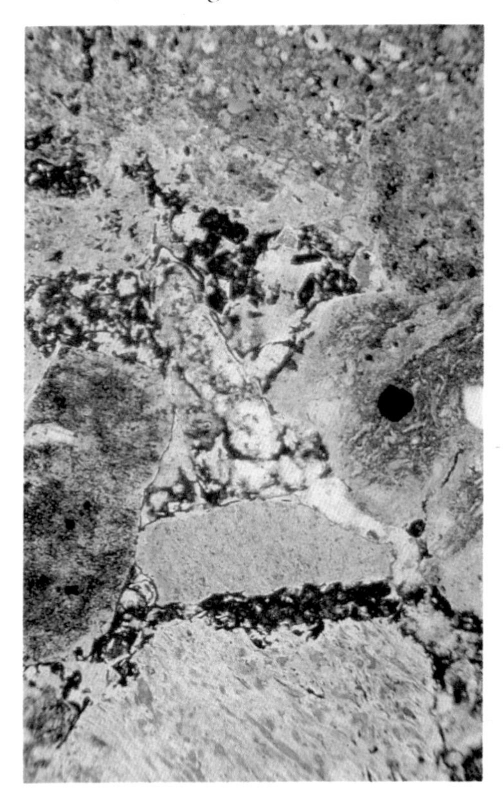

SANDSTONE

Description: Sandstones are essentially products of weathering and the transport of rock particles. They are composed in the main of rock fragments, quartz and feldspar. Geologists describe these fragments as grains or clasts and classify sandstones on their average grain size. Fine-grained sandstone has particles ranging between 0.125 mm (0.005 in) and 0.25 mm (0.01 in) whereas those of medium- and coarse-grained rocks, 0.25 mm–0.5 mm (0.01–0.02 in) and 0.5 mm-2 mm (0.02 – 0.08 in) respectively. Individual grains can vary from well-rounded to angular in shape. This is a good indication of the maturity of the sediments; the more rounded grains having been transported further and longer, or perhaps more often, prior to deposition. Thus sandstones with a predominance of rounded and well-rounded grains are termed mature; immature sandstones possess mostly angular grains, which have only suffered mineral transport by water or wind. Quartz is the most common mineral found in sandstones. It is a hard, resistant silicate mineral which is derived from other rocks and mineral veins. Very clear quartz grains may indicate a volcanic origin, whereas milky-white clasts are usually from veins and fractures. In contrast to quartz, feldspars are softer and less stable chemically. These characteristics make them less resistant to transport and weathering and they commonly break down to form clay minerals. Sandstones vary greatly in terms of colour and hardness. The presence of iron oxides can give a rusty hue whereas cobalt, nickel and titanium may result in richer colourings. Hardness depends greatly on the post-depositional history of the sediment.

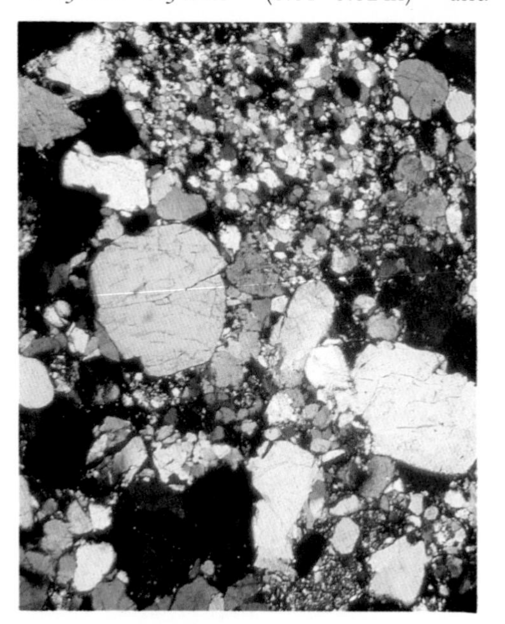

General remarks: Accessory minerals such as glauconite, phosphate and mica together with fossil remains can provide data on the origin and environments in which sandstones were deposited. Glauconite will give the sediment a green coloration, hence the name greensand, applied to glauconite-rich sediments, whereas mica will sparkle and cause the rock to split along distinct planes.
Classification: Clastic sedimentary, rock.
Locations: Worldwide.
Importance/Commercial use: Bulk resource, building stone.

Description: Schists are invariably produced by the metamorphism of mudrocks. They are medium- to coarse-grained rocks, the grain size effectively increasing with increased temperatures and pressures. A traverse across an area of metamorphosed fine-grained sedimentary rocks could show schists grading into phyllites and then into slates. The schists, by definition, would possess higher temperature and pressure minerals. They would also be characterized by an obvious layered arrangement of the constituent minerals. These layers are parallel or subparallel and the rocks split along these well-defined surfaces. In slates these well defined, parallel surfaces along which the rock can be easily split are called cleavage planes. Individual schists are named after their predominant mineral. In mica schists, biotite or muscovite are abundant. Where muscovite is prominent, the rocks are relatively pale coloured and lustrous in appearance. Biotite will also give rise to a lustrous sheen but biotite schists are dark-coloured. Quartz and biotite together produce a distinct banding and usually result in a coarse-grained texture. Deep-red almandine garnets are commonly found in biotite schists. This is particularly true in mountainous regions such as the Alps, Himalayas or Rockies. Such areas have inevitably experienced intense regional metamorphism and schists and gneisses are common rock types produced by such metamorphism. Garnets are usually indicative of higher temperature and pressure regimes. Staurolite and kyanite schists are formed at still higher levels of metamorphism.

General remarks: Tourmaline, zircon and apatite can occur as secondary minerals in schists. The garnets in schists may be sought as semi-precious gemstones.

Classification: Medium to coarse-grained metamorphic rock.

Locations: Worldwide, but associated with areas of mountain building.

Importance/Commercial use: Limited, but may be used as bulk fill material or as source of semi-precious stones.

SYENITE

Description: Syenites are medium- to coarse-grained igneous rocks. They are not common, but are usually found in association with granites or as dykes and sills. The majority of syenites are composed of pale-coloured minerals and are thus termed leucocratic. The opposite term applied to igneous rocks consisting of dark-coloured minerals is melanocratic. Feldspars, particularly alkali feldspars, are the most common mineral in syenites. In some samples, feldspars account for 90 per cent of the rock. Most syenites contain both orthoclase and plagioclase feldspar. Quartz is not common in these rocks, usually being less than 2–3 per cent. Biotite, hornblende, olivine and sodium/potassium silicates (feldspathoids) may occur as accessory minersal or in low percentages. Feldspathoids and quartz never occur together in the same igneous rock. Should quartz represent over 10 per cent of the rock, the term quartz-syenite is applied. Pink or creamy white feldspars are usually the largest crystals observed in syenites. They can be seen with a hand lens or sometimes with the naked eye. In thin section, the various feldspars often can be seen to be intergrown to form perthite and a perthitic texture. Biotite and other dark minerals give syenites a sparkle or lustre. One of the most popular facing stones used in the construction industry is larvikite. It is found near Larvik in Norway and polished slabs adorn many major buildings. The attraction is due to the arrangement and colour of the large feldspar crystals, which when polished, given a blue metallic sheen, a play of light which is termed the Schiller Effect.

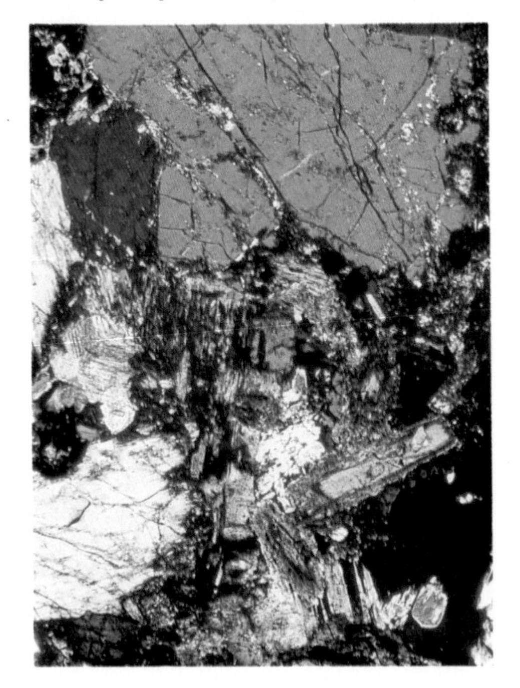

General remarks: Syenites are crystalline rocks which usually have an even grain size. Less commonly they can be found to have large feldspar crystals floating in a fine grained matrix. This rock is termed a rhomb porphyry.

Classification: Medium to coarse-grained crystalline igneous rock.

Locations: Worldwide, but best examples are from Aswan in Egypt and the Bearpaw Mountains of Montana, U.S.A.

Description: This mineral occurs in prisms of variable lengths. These are frequently striated and have a vitreous lustre. The largest crystals of beryl are associated with pegmatitic igneous rocks, with some crystals exceeding 5 metres in length. Individual crystals may be colourless or translucent, but green, yellow blue, pink and red varieties are recorded. Single crystals or masses of 100 tonnes indicate the relative abundance of this mineral in different settings. The larger masses are a source of the element beryllium, whereas small transparent crystals are valuable as gemstones. Dark green emeralds and pale blue aquamarine are varieties of beryl and are examples of the commerical value of this mineral. Beryl is an extremely hard mineral that cannot be scratched with a penknife. It has a glassy, conchoidal fracture and a white streak. Crystals of beryl exhibit a sixfold symmetry with three horizontal axes of equal length and one larger vertical axis. The crystals are referred to the hexagonal crystal system. Beryl is found in association with igneous and metamorphic rocks. It may also occur as a vein mineral, as a result of hydrothermal activity. Beryllium, aluminium, silicon and oxygen occur within the molecular structure of this mineral. Like tourmaline, it is a ring silicate, a rarer variety of the most common rock-forming silicate minerals.

General remarks: Beryl in various forms provides us with examples of some of the largest and most prized crystals known to man. Chromium is the element responsible for the green colour in emeralds, whilst traces of iron provide the shades of blue and blue green seen in aquamarine.

Classification: Hexagonal, crystal system ring silicate.

Locations: Worldwide, but emeralds of gem quality are confined to a number of well known localities such as the Muzo region of Colombia, South America.

Importance/Commercial use: As source of beryllium and as gemstones of different qualities.

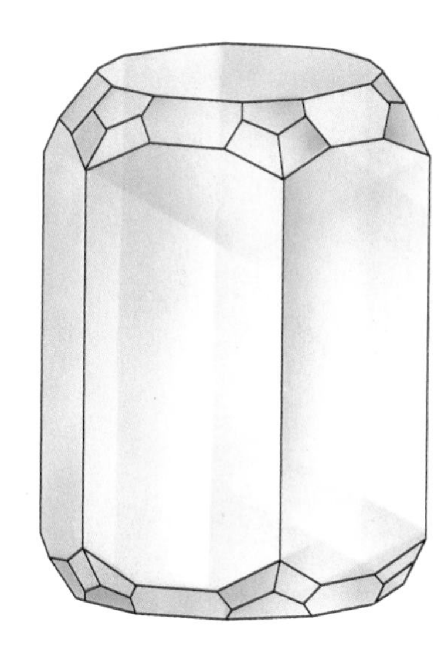

BIOTITE

Description: Biotite is a mica and is one of several rock-forming minerals that cleave perfectly along horizontal planes. It is black or brown in colour with a lustrous sheen. Geologists frequently refer to 'books' of biotite, or other micas, in their descriptions of igneous or metamorphic rocks. This provides an apt comparison, as the layers of mica peel open like the pages of a book. Biotite is composed of potassium, magnesium, silicon, and aluminium. The magnesium may be completely replaced by ferrous or ferric iron. Micas have three crystallographic axes of unequal length with two of these axes at right angles to each other. Such crystals belong to the monoclinic crystal system. Orthoclase, gypsum and epidote are other minerals whose crystals belong to this system. In hand specimen, biotite may occur as 'books' (lamellar aggregates) or as tabular crystals. Small flakes or platelets and short prisms may also be observed where the mineral is simply an accessory. Although generally black or brown, thin flakes of biotite can be greenish or even translucent in colour. In this case, care must be taken in the identification of biotite as against its sister mineral phlogopite. Fortunately, the latter has a distinctive coppery appearance which aids its identification; it also occurs more commonly in metamorphosed limestones and dolomites and magnesium-rich igneous rocks. Biotite is found in granites, diorites, schists and gneisses.

General remarks: Biotite can occur as single flakes or as aggregrates of considerable size. Its hardness, on a scale 1 to 10 in which diamond is 10, is between 2 and 3. It may just be scratched by a fingernail.

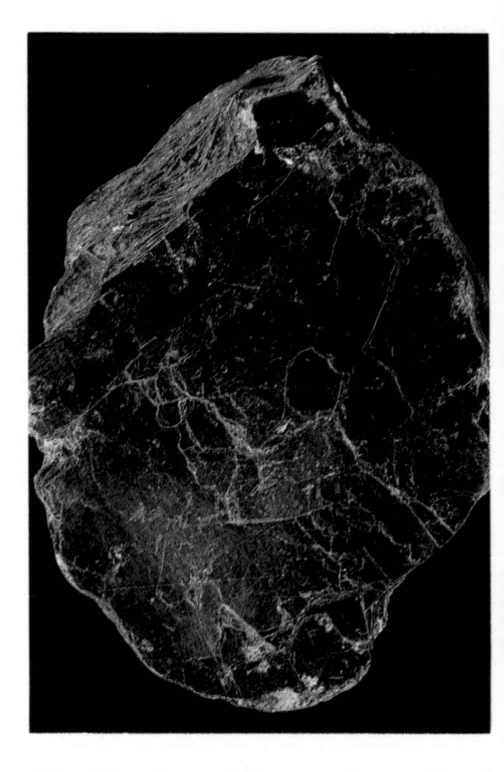

Classification: Mica or sheet silicate. Monoclinic crystal system.
Locations: Worldwide, mainly in igneous and metamorphic rocks.
Importance/Commercial use: Biotite, unlike muscovite and phlogopite (two minerals used in the electrical industry), has little commercial value.

CALCITE

Description: Calcite is one of the most important rock-forming minerals. It is a carbonate, the formula of which is $CaCO_3$. Apart from the silicates (e.g., quartz and feldspar), carbonates are the other major group of rock-forming minerals. Calcite may be recognized by its white to transparent appearance, vitreous lustre and white streak. It is relatively soft and is easily scratched with a penknife. Probably the simplest way in which geologists identify calcite is by using dilute hydrochloric acid. A few drops of this dilute acid act on calcite to produce a vigorous fizzing reaction or effervescence, which is easily observed. Crystals of calcite are frequently found in limestone cavities, fractures and veins. They can be sharply pointed (scalenohedra) or low-faced and flat (rhombohedra). Of these, the first are referred to as 'dog-tooth' calcite, the second as 'nail-head' calcite. The crystals belong to the trigonal crystal system. They are characterized by four axes of symmetry; the three horizonal axes are equal in length to each other, but are shorter than the vertical axis. Elongate prisms are a common crystal form within this system. Calcite exhibits three perfect sets of cleavages; excellent rhombohedra can be obtained by gently tapping pieces off a larger block. Many animals and some algae use calcite to build skeletons and to support internal structures. Fibrous and lamellar calcite are found in the shells of clams and shellfish. Calcite also replaces the less stable mineral aragonite over time under normal temperatures and pressures. In metamorphic terrains, calcite is the predominant mineral in marble. Dolomite replaces calcite in certain limestones.

General remarks: Calcite is often found in association with metallic ore veins. It is also found as stalactites and stalagmites in caves and as travertine and tufa in hot or cold water springs.
Classification: Carbonate mineral of the trigonal crystal system.
Locations: Worldwide in limestones, fractures and veins.
Importance/Commercial use: As a flux in smelting and as a fertiliser.

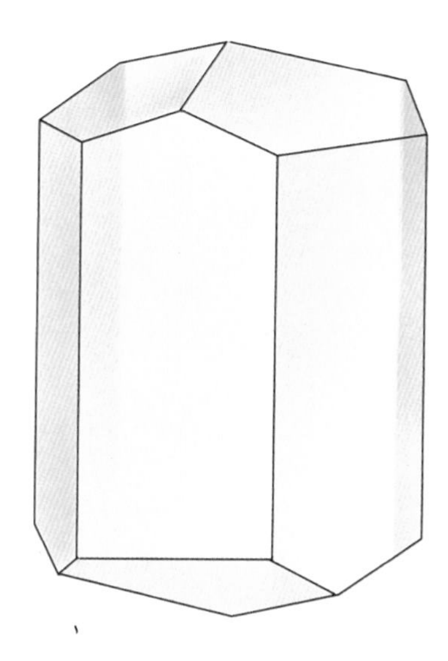

CHALCEDONY

Description: Chalcedony or chalcedonic silica is a variety of quartz. It usually occurs as rounded masses, termed mamillated or botryoidal, or as stalactites. These forms of chalcedony are found in cavities, vugs or fissures. Unlike quartz, it tends to have a fibrous rather than a granular texture. It also has a wide range of colours, with the more exotic colours present in semi-precious varieties. Carnelian is red, as is jasper; chrysoprase is bright green and bloodstone green with red spots. Chalcedony has a rather waxy appearance and a conchoidal or glassy fracture. It is hard, like quartz, but can be scratched with a penknife. It is described as cryptocrystalline, a term which infers that the crystals can only be observed under a microscope. Fluids or solutions rich in silica are the source of chalcedony in cavities and fissures. In volcanic areas, silica-bearing solutions can replace the shell structure of fossils, with chalcedony replacing calcite or aragonite. Chalcedony differs from agate, another form of quartz, in that it lacks the obvious banding or layering of agate. Agate is often associated with volcanic areas. It is commonly found as a lining in geodes. This material is again sliced and polished for sale as an ornamental or semi-precious stone. Both chalcedony and agate occur as vug infills in lavas. Chalcedony may also impregnate the tissues of fossil wood. These are extremely attractive and many are sliced to form semi-precious ornaments.

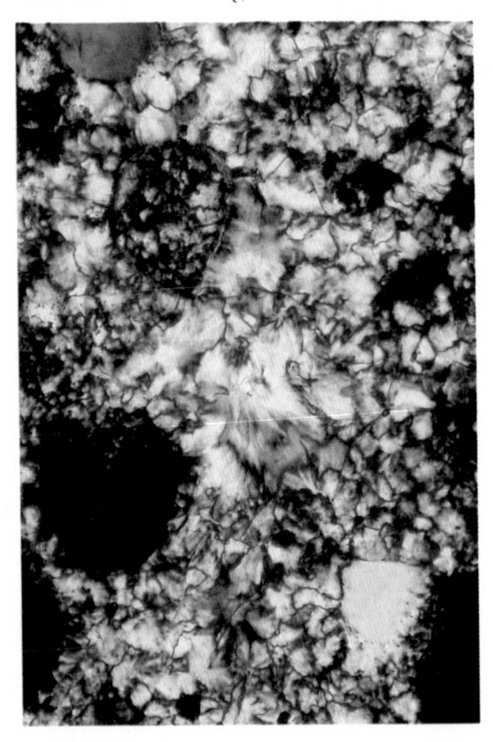

General remarks: The less precious forms of chalcedony are transparent, white or grey. It has a white streak and is difficult to separate from botryoidal quartz deposited on the sea-floor.

Classification: Cryptocrystalline form of quartz.
Locations: Worldwide.
Importance/Commercial use: Semi-precious and ornamental value.

FLUORSPAR (Fluorite)

Description: This transparent, purple blue, yellow, green or brown mineral is often found in association with quartz and barytes. It usually takes the form of cubic crystals or granules. The transparent crystals may be mistaken for calcite, but fluorspar does not react with hydrochloric acid. Fluorspar has a white streak and a vitreous lustre. If fractured, it tends to splinter or break with curved faces (conchoidal fracture). It is moderately hard but can be scratched with a copper coin. As a member of the cubic crystal system, fluorspar has three equal axes all at right angles to each other. The cube is the essential crystal form of this group, but fluorspar crystals are characteristically modified cubes. These have additional faces along the leading edges or at each corner. In some cases the cubes may appear to grow out of each other, with one crystal appearing to bear a small pyramid on each face. This feature is termed twinning, and the form described is called an interpenetration twin. The various colours observed for fluorspar are due to impurities. In the Peak District of Derbyshire, a blue-banded variety of fluorspar is associated with mining. The local name for these minerals is Blue John, and perfect samples are prized by collectors. The Peak District is a limestone area and the Blue John is associated with hydrothermal veins. It may be found in association with tin-bearing veins. These are mostly associated with granites and are produced by hot gases rich in fluorine, passing through fissures and fractures.

General remarks: Although fluorspar does not react with hydrochloric acid, it is used in the manufacture of the same fluid.
Classification: Calcium fluoride: cubic crystal system.
Locations: Worldwide, but mostly in hydrothermal or pneumatolytic veins.
Importance/Commercial use: Transparent crystals used in optical industry. Rich deposits quarried as flux for steel manufacture.

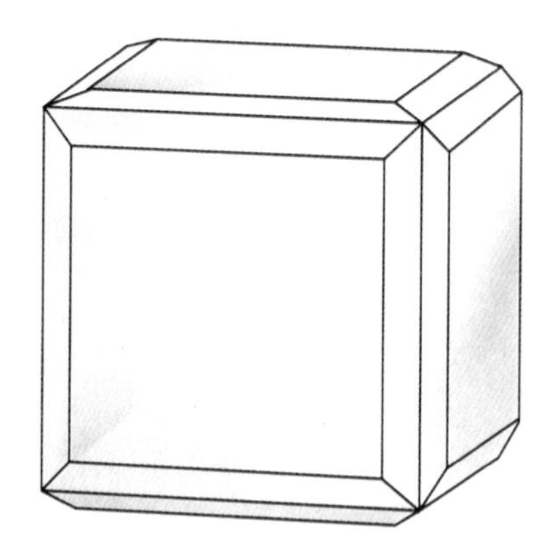

GALENA

Description: Galena is lead sulphide, the formula of which is PbS. Like fluorspar, it is a member of the cubic crystal system and the common crystals are mainly cubes. These may be modified with a small triangular face on each corner; this is termed a cube octahedron. Galena is relatively soft and can be scratched with a finger nail. As one would expect with a lead-based mineral, it has a specific gravity of 7.8. Specimens of, or containing, galena feel heavy when picked up. Galena also has a well-developed metallic lustre; this, coupled with its density, makes galena relatively easy to identify. If tapped, the mineral cleaves into cubes, although massive pieces are brittle and may give a stepped or conchoidal fracture. The lead gives rise to a grey streak. Galena is often found as a vein mineral. It is associated with sulphide bodies that are formed by hydrothermal activity and is often found with blende or sphalerite. The latter is also cubic but has a brown-yellow or white streak. Together, the two minerals are worthy of further economic investigation, as both are important ores. Galena is a major lead ore and sphalerite a major source of zinc. Both are found in areas affected by metasomatism. This process is induced by either hot liquids or gases passing through a rock, following igneous intrusion. It involves either an addition of minerals or their possible removal. The process is perhaps a more gentle form of metamorphism. Galena and blende are also found in sedimentary sequences. Galena and blende are also found in sedimentary sequences. An aggregate of quartz, galena and blende makes a fine collectors item.

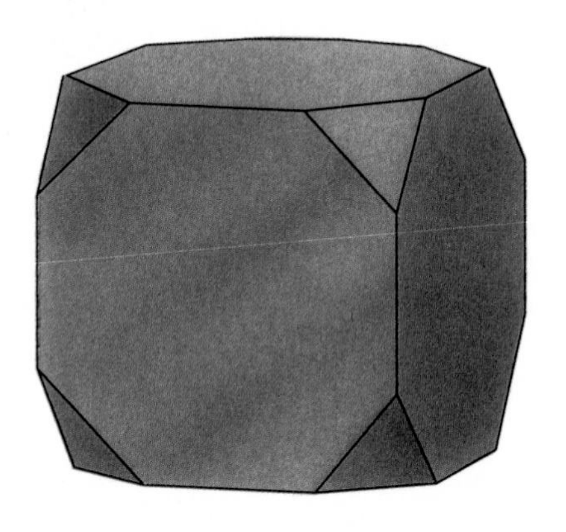

General remarks: Minerals found in association with galena and blende are often of little or no value commercially. Fluorite, quartz, barytes and calcite are such minerals; they are termed gangue minerals by miners.

Crystals of galena in association with calcite crystals.

Classification: Load sulphide: cubic crystal system.
Locations: Worldwide.
Importance/Commercial use: Lead ore.

Description: Gemstones are minerals that possess beauty, durability and have a rarity that makes them desirable to men and women alike. Most gemstones are harder than quartz (7) and cannot be scratched by the blade of a penknife. Some are brittle or are affected by heat and are therefore of limited use. The test of a good gemstone is its resistance to wear and tear. Colour, clarity, the absence of impurities and the absence of cleavage are also critical to the final assessment of a crystal to be of gemstone quality. The complete absence of colour in diamonds, the green of an emerald or the rich red of a ruby are desirable colour characteristics. Cutting by a skilled craftsman can further enhance the beauty of a stone by bringing out hidden optical properties. The cut is effected with the minimum loss of weight to an individual gemstone, but facet can bring out internal reflections. Rarity is perhaps the most important factor controlling the value of a gemstone, although some extremely rare minerals are poorly-coloured and lack commercial value. Many gemstones are found in stream deposits. They are resistant to erosion and may have been carried many miles from their parent rock. Most gemstones are associated with pegmatite veins which were formed during the latter stages of the instrusions of granitic bodies. Others such as rubies and sapphires are associated with basaltic lavas, whilst diamonds are associated with pipes of basic igneous rock called kimberlite. The gemstone opal, is formed by the slow precipitation of silica in a variety of environments.

General remarks: Gemstones are drawn from a number of mineral groups. Diamonds are pure carbon whereas rubies are aluminium hydroxides.

Classification: Minerals mostly associated with igneous rocks, rarely sedimentary or organic.

Locations: Worldwide, but mostly in areas with dyke intrusions.

Importance/Commercial use: Of considerable importance to industry and to the world of haute couture.

Left: Garnet.
Below: Topaz varieties.

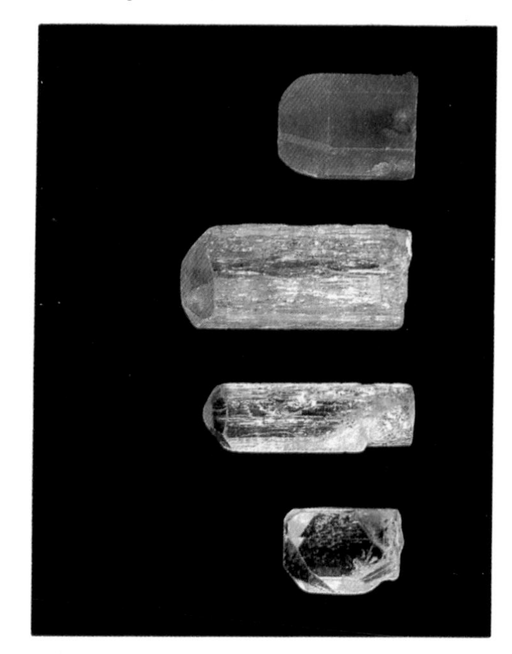

HAEMATITE (hematite)

Description: Haematite is an oxide of iron and belongs to the trigonal system. It is commonly found as massive ore deposits and compact aggregates. The most handsome finds are the botryoidal or mammilated masses known as kidney ore. These are usually steel gray to black in colour, but some masses of kidney ore can be a rich red colour. The streak is characteristically dark red and the lustre metallic. As you would expect of an iron ore, haematite is relatively hard and can only be scratched with a penknife. It has an uneven or subconchoidal fracture and not, surprisingly, is brittle. Individual crystals can be fibrous and groups of crystals can form distinct rosettes. Haematite is slightly heavier than pyrite but not as heavy as galena. It is one of the major sources of iron ore, with the major deposits associated with sedimentary successions. In igneous rocks, it is usually found as an accessory mineral. Elsewhere, haematite is found as a product of hydrothermal or volcanic activity and is also found in metamorphic sediments. The largest deposits of this ore mineral are mined around Lake Superior, Quebec, Canada and in Brazil. The iron content of the ore is over 70 per cent. Crushed haematite is known as red ochre and is used as a pigment. Although it is an iron ore, haematite is not normally magnetic. The reddish colour of some feldspars is due to haematite.

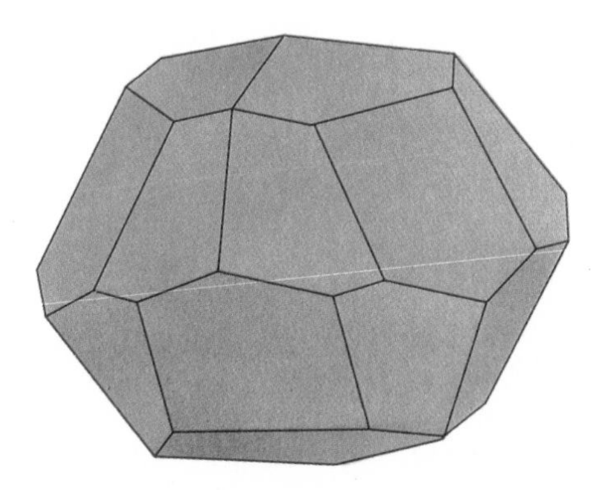

General remarks: Haematite, magnetite and limonite are three important iron ores. In contrast to haematite and magnetite, limonite is more variable in composition, containing hydroxides. It is a product of oxidation under normal temperatures and pressures and is often responsible for the brown staining observed in rocks or on rock surfaces.

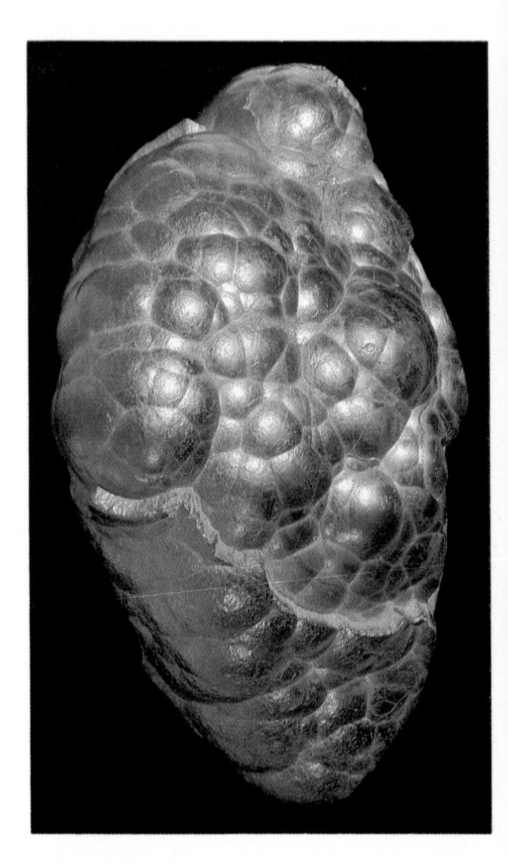

Classification: Iron oxide, cubic crystal system.
Locations: Worldwide.
Importance/Commercial use: Iron ore; red ochre used as a pigment.

HORNBLENDE

Description: Hornblende is a monoclinic mineral normally found as long or short prisms that are frequently six-sided in cross section. It can also occur as aggregates or as an accessory mineral. Hornblende is the commonest amphibole, a group of silicate minerals whose atomic structure comprises a double chain or ribbon structure of silica tetrahedra. Innumerable numbers of these double chains are arranged parallel to each other, running along the long axis of each crystal. The arrangement of these ribbons is responsible for the intersection of two cleavage planes at 120 degrees to each other. Amphiboles are rather difficult to identify, but the combination of cleavage and colour is usually sufficient. Hornblende is light green to dark green to almost black in colour, with a vitreous lustre. It is also relatively hard; the surface is resistant to a copper coin but will be scratched by the blade of a penknife. Hornblende is an important rock-forming mineral and is found in many types of igneous rock. It is most common in acidic and intermediate igneous rocks such as granodiorites and diorites, but is also found in gabbros. These are basic rocks noted for the presence of calcic plagioclase feldspars. Hornblende is also present in syenites, and metamorphic rocks such as schists are also characterized by the presence of hornblende. In medium-grade, regionally-metamorphosed regions, hornblende schists are a typical rock. Amphiboles are abundant in amphibolites, or regionally-metamorphosed basalts and dolerites.

General remarks: Hornblende, pyroxene, olivine and biotite are dark-coloured or mafic minerals. The proportion of light to dark-coloured minerals within an igneous rock enable geologists to erect a scheme of classification based on colour, which is best applied to medium-and coarse-grained igneous rocks.
Classification: Amphibole; monoclinic crystal system.
Locations: Worldwide.
Importance/Commercial use: Of no commercial value.

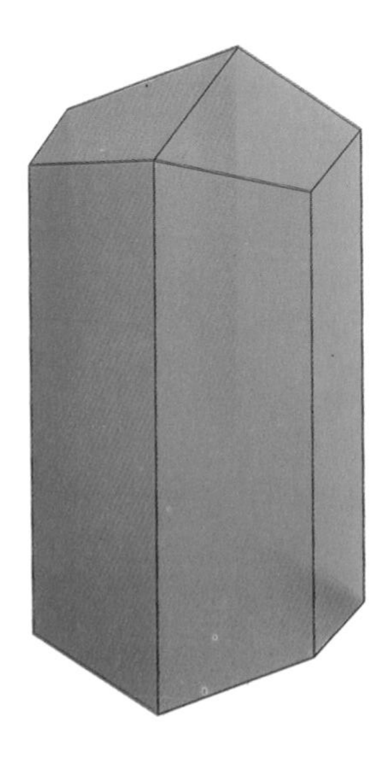

KYANITE

Description: Kyanite, like sillimanite and andalusite, is an aluminium silicate. Kyanite, however, belongs to the triclinic crystal system, while the other two are are orthorhombic. Crystals of each of these three minerals have a prismatic habit, but kyanite crystals can also be flat or bladed and can occur in bladed aggregates. Strangely, the crystals are flexible and may occur in bent or twisted form. Another characteristic of this pearly or vitreous mineral is that its hardness varies along the length and across the width of each crystal. Disthene, one of the old names for kyanite is from the Greek meaning two kinds of hardness. Generally, the crystal may be scratched with a penknife. The colour varies from translucent through white to green and blue, a patchy blue being the most common colour. Unfortunately, the crystals are often brown stained, due to weathering. Whatever the colour, the streak is white. Kyanite crystals cleave well. This mineral is found in medium grade metamorphic rocks and the name is used as a prefix for schists and gneisses. Kyanite schists are of a higher grade than garnet-mica schists, but are formed at lower temperatures than sillimanite-rich rocks of the same genre. Sillimanite is a high temperature/pressure polymorph of kyanite. Perfect blue-bladed crystals are recorded from the kyanite schists of the Hurricane Mountains, North Carolina, United States of America. Well-formed crystals can be found in pegmatites and quartz veins.

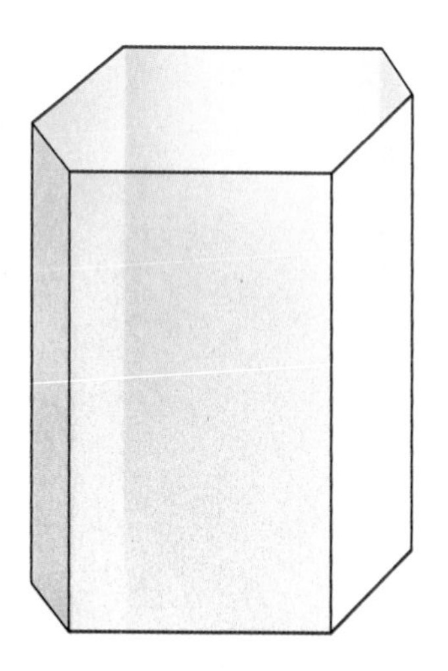

General remarks: Of the three aluminium silicates noted above, kyanite and andalusite can be heated to 1300°C to produce mullite, a silicate used in the manufacture of spark plugs.

Classification: Aluminium silicate: triclinic crystal system.
Locations: Worldwide.
Importance/Commercial use: Production of mullite. Andalusite and sillimanite cut as gemstones.

Description: Olivine is a silicate mineral of either iron or magnesium. It has a relatively simple atomic structure and is referred to the orthorhombic crystal system. Well-formed crystals of olivine are uncommon, the mineral usually occurring as irregular crystals in igneous rocks or as granular aggregrates. Of the two major varieties of olivine, forsterite, which is magnesium-rich, is usually green in colour whereas fayalite, with iron, is brown to black. Both have a white or grey streak. Olivine is almost as hard as quartz and can be scratched with a penknife. It has a vitreous (glass-like) lustre and a poorly-defined cleavage. The fracture is conchoidal, as seen in volcanic glasses. Olivine is an important rock-forming mineral. It is usually found in basic and ultrabasic rocks which have little or no quartz. These include basalts, gabbros and the dark-coloured, ultrabasic peridotites. In the latter, olivine may be the sole mineral. Such rocks are termed dunites or olivinities. Not surprisingly, dunites are green-coloured with a granular texture. Olivine basalts are commonplace with crystals or nodules of the green lustrous mineral as a characteristic feature. Stony-iron meteorites or pallasites are also rich in olivine. Peridot is the name given to transparent olivines of good colour, which are cut and used as gemstones.

General remarks: Although olivine is a common rock-forming mineral, it is unfortunately of little economic value. Fayalite, however, does have a very high melting point and can be used in the manufacture of high temperature bricks.

Classification: Magnesium or iron silicate; orthorhombic crystal system.

Locations: Worldwide, in basic and ultrabasic rocks.

Importance/Commercial use: Peridot as gemstones, fayalite in the manufacture of refractory bricks.

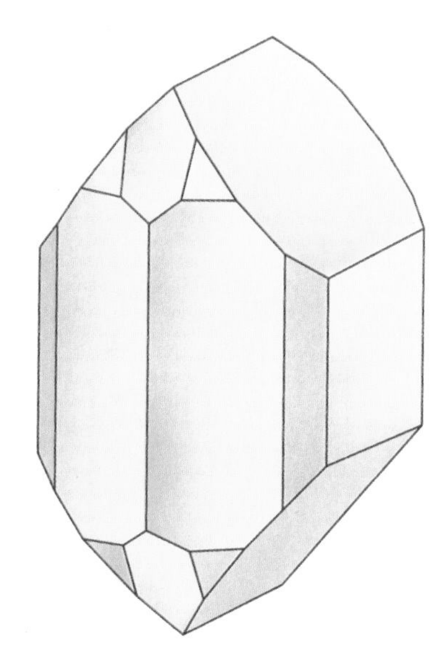

OPAL

Description: Opal is a silicate mineral with a cryptocrystalline structure. It is a hydrous variety of cristobalite; this means that water is present in the molecular structure. Cristobalite is a variety of quartz. Opal is characterized by its vitreous or resinous lustre and low specific gravity. It has a conchoidal fracture and may be scratched with a penknife. The colour may vary considerably from milky white to black, with red, brown, green and blue stones also being recorded. The most sought after opals are those with an iridescence. This is due to the splitting of light by the internal molecular structure of the mineral and not to impurities as in other minerals. Colourless, non-precious, opals have no internal reflections. Opal may occur in a massive form, which may or may not be botryoidal. Stalactitic opal is also recorded, together with more normal vein-associated material. In areas of volcanic and hydrothermal activity, opal or opaline silica replaces other substances such as wood and calcite. Polished sections of opalized wood and opaline fossils of sea shells are prized collectors items. Opaline silica is also found as the skeletal material of microscopic marine organisms. These float in the open oceans and on death, sink to the sea bed to form a siliceous ooze. In the geological record, such deposits may ultimately give rise to cherts; they are a good indication of deep water environments.

General remarks: Hot springs and geysers are known centres for the production of opaline silica. Waters rich in silica can be traced back to a magma chamber. At depth, the silica will fill or line fissures or fractures. At the surface, it will take the form of massive aggregates or stalactites, or replace the skeletal structure of plants and animals.

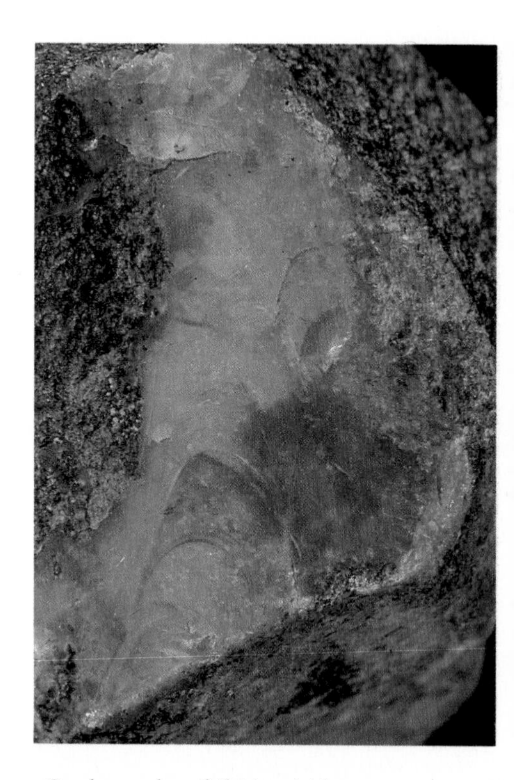

Opal sample exhibiting iridesence.

Classification: Hydrous silicate.
Locations: Worldwide; gemstones from Australia and Africa.
Importance/Commercial use: As precious stone; diatom-rich opaline sediments used as abrasive or filler materials.

ORTHOCLASE

Description: Orthoclase is one of the most common feldspars found in igneous, metamorphic and sedimentary rocks. It is a potassium feldspar with a chemical formula of $KAlSi_3O_8$. Orthoclase crystals are commonplace in acid igneous rocks, such as granites, where they are usually flattened prisms. Twinning is a common occurrence. Orthoclase is white to red, with a flesh pink colour being characteristic of many crystals. It has two well-developed cleavages and an uneven or conchoidal fracture. Most feldspars have a hardness value of between 6 and 7, which means they can be marked with a penknife. The streak of orthoclase is white and most crystals exhibit a vitreous or pearly lustre. Both orthoclase and sanidine, a high temperature variety, are monoclinic, whereas microcline, another variety of potassiuum feldspar, is triclinic. It is difficult in hand specimen to distinguish between orthoclase and microcline. A rather suspect guideline is that orthoclase occurs in granites and microgranites and microline is found in pegmatites and vein deposits. Microcline is sometimes bright green in colour; this variety is called amazonstone and it is seen as a cut and polished ornamental stone. Microcline may also be used in the manufacture of enamel, glass and porcelain. In contrast, orthoclase is of little or no commercial value. However, orthoclase is susceptible to alteration from the action of hydrothermal solutions and from the normal processes of weathering. A common alteration product is kaolin, which is the main mineral constituent of china clay, which is mined for this purpose in Devon and Cornwall, England.

General remarks: The Carlsbad twin of orthoclase is an excellent example of a penetration twin, where two portions of the same mineral are intergrown.
Classification: Potassium feldspar; monoclinic crystal system.
Locations: Worldwide; excellent crystals found in the Shap Granite from the Lake District in England.
Importance/Commercial use: Of no commercial value, except for kaolin.

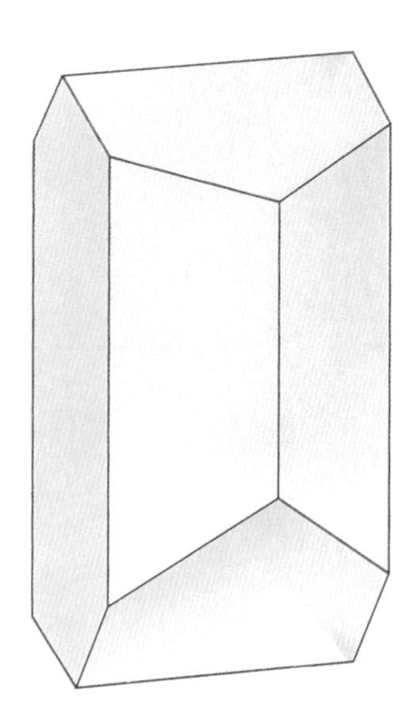

PLAGIOCLASE

Description: Plagioclase feldspars are the sodium and calcium-rich equivalents of the potassium feldspars such as orthoclase. Plagioclase feldspar belongs to the triclinic crystal system. It occurs as prisms, tabular crystals or in massive granular aggregrates.

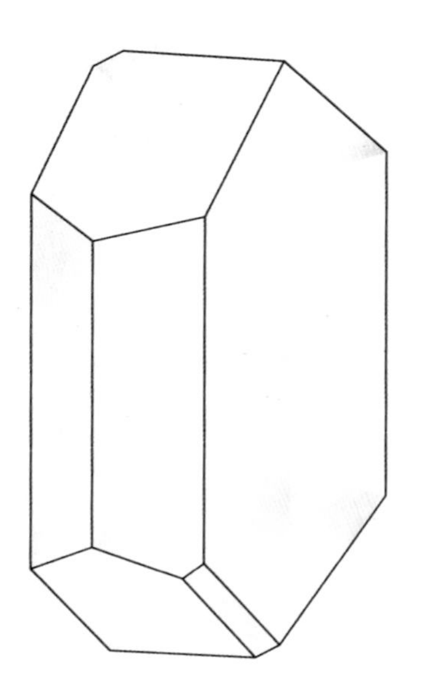

Twinned crystals are common and repeated twinning, where portions of the mineral appear to be stacked on end, is a characteristic feature of plagioclase feldspars. However, such repeated twinning is often difficult to see with the naked eye and is best observed under the polarizing microscope. Plagioclase exhibits two good cleavages and an uneven fracture. The colour varies from white to pink, green and brown. Transparent or translucent crystals also occur. They have a white streak and a vitreous lustre. The chemistry of plagioclase feldspars changes gradually from sodium rich-to calcium-rich, with albite, oligoclase, andesine, labradorite, bytownite and anorthite representing a transitional suite of minerals. Labradorite often exhibits a blue-green sheen or play of colours on cleavage surfaces. The plagioclase feldspars are important rock-forming minerals. They are particularly abundant in igneous rocks and are used in the subdivision of acidic, intermediate, basic and ultrabasic igneous rocks. Sodium-rich plagioclase is typical of granitic rocks, whilst calcium plagioclase is characteristic of basalts and gabbros. Anorthosites, medium to coarse-grained ultrabasic rocks, consist of masses of labradorite, bytownite and anorthite. Plagioclase may account for over 90 per cent of certain igneous rocks.

Albite crystals in association with quartz crystals.

General remarks: Plagioclase feldspars may also occur in pegmatites (albite) and in metamorphic and sedimentary rocks.
Classification: Sodium-calcium feldspar: triclinic crystal system.
Locations: Worldwide
Importance/Commercial use: Albite and oligoclase have important uses in the production of ceramics.

Above: Native silver.
Right: Native gold.

Description: Economic minerals are essentially those which can be extracted in considerable quantities at a price consistent with demand. Bulk materials include coal and limestone but economic minerals, which occur in smaller amounts, have a greater monetary value. Metals such as lead, zinc, copper and iron are correctly described as economic minerals. They can, however, be mined in great quantities and could hardly be described as precious. Gold and silver however, are relatively scarce and the fact that gold does not tarnish has led to its description as the 'King of metals'. Both gold and silver are maleable and ductile, that is, they can be shaped without fracturing and are thus very important metals in the jewellery trade. Gold is yellow in colour with a metallic lustre. It is found in association with quartz and pyrite in hydrothermal veins. Gold and silver are termed 'native elements', meaning that they naturally occur in their pure form, rather than as mineral compounds. Following mining, these minerals are broken down and smelted to remove the metallic component. Gold is resistant to chemical action and once freed from the vein material, it may be transported by water and found amongst alluvial deposits. Silver, gold and platinum are cubic minerals. Silver is also associated with hydrothermal veins and more rarely occurs in placer deposits. It has a metallic lustre and a bright silver-white streak. Platinum is found in association with ultrabasic rocks; it has a steel-grey streak. Like gold and silver, platinum is malleable. Platinum is resistant to acids and is used in chemical laboratories worldwide.

General remarks: Both gold and silver were previously used in the manufacture of coins and gold was formerly the currency standard.

Classification: Gold and silver are native elements, cubic system.

Locations: Worldwide in areas of igneous activity.

Importance/Commercial use: Gold silver and platinum are important to the jewellery trades; gold is used in scientific and electrical apparatus; silver is extensively used in the manufacture of photographic materials.

PYRITE

Description: Pyrite is an iron sulphide. It is often found as cubes, but pyritohedra and octahedra also occur. Aggregates of striated cubes make for beautiful ornaments. Excellent samples of such aggregates are found in Northern Seporades, Greece.Pyrite may also occur as nodules or as pseudomorphs after salt crystals. It is also a common replacement mineral which impregnates and replaces wood or bone tissue and the shells of molluscs. The mineral is a brassy-yellow colour which can tarnish to a purplish iridescence. Drawn across an unglazed ceramic plate, pyrite leaves a brown-black streak. Pyrite is known as 'fools gold' due to the pale yellow coloration and metallic lustre. It is, however, much harder than real gold and has a much lower specific gravity (5 against 19.3 for gold). Pyrite is often associated with hydrothermal veins and is also found in both metamorphic and sedimentary rocks. It is the most common sulphide in many regions and environments. Marcasite, which is also an iron sulphide, forms at lower temperatures than pyrite and frequently occurs as nodules with a radiating structure. Such nodules are commonly found in chalk sediments. These have a brittle fracture and a yellow-brown exterior. Chalcopyrite or copper pyrites has a slightly darker more yellow colour. It is softer than pyrite with a lower specific gravity. Chalcopyrite is a copper iron sulphide and is a major copper ore. Unlike gold, both forms of pyrite are soluble in nitric acid and are not malleable.

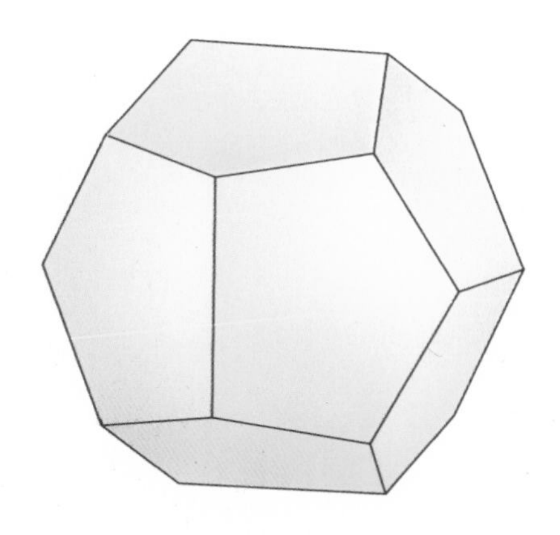

General remarks: Iron pyrites is an attractive mineral, but it weathers badly and decomposes over a relatively short period of time if left untreated.

Classification: Iron sulphide: cubic crystal system.
Locations: Worldwide in igneous, metamorphic and sedimentary rocks.
Importance/Commercial use: Exploited along with chalcopyrite in the production of copper. Sulphur component used in the manufacture of sulphuric acid.

Description: Quartz is one of the purest silicates. It is silicon dioxide that contains only traces of other elements. Six-sided prisms with six faces at either end are the characteristic crystals of quartz, although perfect crystals are relatively rare and amorphous masses predominate. Quartz belongs to the trigonal system. Many crystals are twinned, with penetration and contact forms respectively termed Brazil and Japan twins. Crystals may just be scratched with a penknife and have a conchoidal fracture. Perfect crystals are usually colourless or white. Impurities can give rise to a number of colours with specific names such as rose quartz, citrine and smoky quartz. Amethyst is the purple-violet variety of quartz. A white streak and a glassy, vitreous lustre are characteristics of this silicate. Quartz is very resistant to weathering and to attack by most acids. It is the most common of all silicates and is present in igneous, metamorphic and sedimentary rocks. Acid igneous rocks have more than 10 per cent visible quartz. The mineral is also abundant in veins where it occurs as a gangue mineral. Through time, the weathering of acid igneous rocks and vein minerals produce detrital quartz grains. These are transported by wind and water and are deposited as sands, which ultimately become indurated to form sandstones, in both marine and non-marine environments. Sandstones composed almost entirely of quartz are termed quartzites. 'Tigers eye' is a pseudomorphous structure, where quartz has replaced fibres of crocidolite. Iron oxides gives 'tigers eye' its characteristic appearance.

General remarks: Quartz sands are used in the construction industry or as a flux in smelting.
Classification: Silicon dioxide: trigonal crystal system.
Locations: Worldwide, in many geological environments.
Importance/Commercial use: As a bulk resource; perfect crystals used in the production of optical instruments.

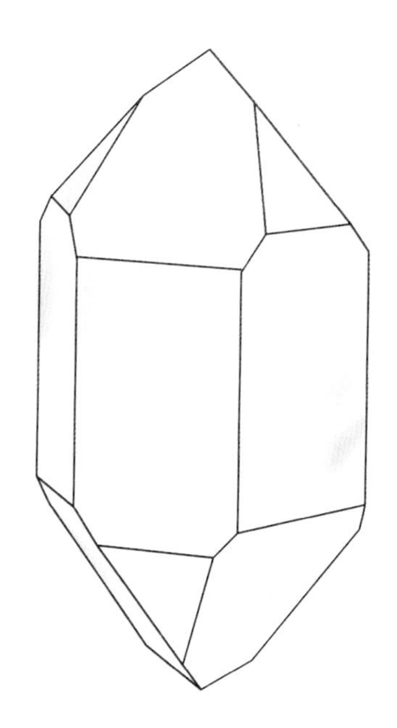

TOURMALINE

Description: A number of minerals are grouped under the heading of tourmaline. They are essentially sodium-rich aluminium silicates, with various percentages of magnesium, boron, iron and lithium. Tourmaline (sensulato) belongs to the trigonal crystal system. It is as hard as quartz and is typically found as elongate prisms. These may be strongly striated with a somewhat rounded cross-section. Tourmaline has a poor to very poor cleavage and an uneven or conchoidal fracture. It has a white streak and a vitreous lustre. Different colours characterize different varieties of tourmaline. Schorl is often black, rubellite pink and red and indicolite blue, among others. Green and colourless varieties are also known. Tourmaline is usually found in pegmatites or in granites that have been affected by the migration of boron-rich fluids. It is a common accessory mineral in schists and gneisses and brown magnesium-rich tourmaline is found in metamorphosed limestones. Delicate needles of greenish black tourmaline can be found in altered granites. The needles radiate outwards from the edges of feldspar crystals. They are embedded in quartz and such clusters are called tourmaline suns. Granitic rocks in which the feldspars have been replaced by quartz and tourmaline are called roche or schorl rocks. Such rocks are associated with the granites of south-west England.

General remarks: Elongate crystals of tourmaline set in a quartz matrix are sought as collectors items. Some of the colours give tourmaline a gemstone quality.

Classification: Sodium aluminium silicate: trigonal crystal system.
Locations: Worldwide; best seen in pegmatite dykes and metasomatized granites.
Importance/Commercial use: Tourmaline has piezoelectric properties and is used in the manufacture of pressure gauges. Limited importance as gemstone.

GLOSSARY

Acidic rocks: Quartz-containing igenous rocks.

Amorphous: Not possessing definite crystal structure.

Amygdaloid: Rocks in which vesicles have become filled with secondary minerals.

Clastic: Rock composed mainly of fragments transported to their place of deposition.

Cleavage: Tendency of some minerals to split in particular directions giving smooth, plane surfaces.

Conchoidal: A variety of fracture with the approximate shape of one half of a bivalve shell.

Cryptocrystalline: Composed of very fine or microscopic crystals.

Detritus: Fragments of rock remaining from the disintegration of older rocks.

Dyke (dike): A tabular rock body, usually igneous, which cuts across the surrounding rock strata.

Foliated: Made up of thin leaves of material.

Gangue: The worthless minerals in an ore.

Groundmass: Crystalline background material for phenocrysts in a porphyritic rock.

Lithification: The process whereby sediments are changed into solid rock.

Matrix: The material in which a specific mineral is embedded.

Metamorphic rock: Rock formed from igneous or sedimentary rock that have been subjected to large changes in temperature, pressure and chemical environment.

Monoclinic: Crystal system with three unequal intersecting axes, two of which are at right angles and the third inclined.

Nodule: Rounded lump of rock or mineral.

Phenocryst: A crystal substantially larger than the crystals of surrounding material.

Phyllite: A metamorphic rock not as well-developed as a foliated schist but in a more advanced state of recrystallization than a slate.

Porphyritic: Igneous rock containing conspicuous phenocrysts in a fine-grained or glassy groundmass.

Rhombohedra: Low-faced, flat crystals.

Scalenohedra: Sharply-pointed crystals.

Triclinic: Crystals characterized by three axes of unequal length and all oblique to one another.

Trigonal: Three angled.

Twinning: Two or more crystals which have intergrown in a definite way.

Vesicles: Small cavities caused by gas expansion.